W9-BDD-504

The Book Of Five Rings

Miyamoto Musashi

MB

ISBN 978-1-60386-832-7

DIGITALIZED BY
WATCHMAKER PUBLISHING

Contents

Preface

I have been many years training in the Way of strategy, called Ni Ten Ichi Ryu, and now I think I will explain it in writing for the first time. It is now during the first ten days of the tenth month in the twentieth year of Kanei (1645). I have climbed mountain Iwato of Higo in Kyushu to pay homage to heaven, pray to Kwannon, and kneel before Buddha. I am a warrior of Harima province, Shinmen Musashi No Kami Fujiwara No Geshin, age sixty years.

From youth my heart has been inclined toward the Way of strategy. My first duel was when I was thirteen, I struck down a strategist of the Shinto school, one Arima Kihei. When I was sixteen I struck down an able strategist, Tadashima Akiyama. When I was twenty-one I went up to the capital and met all manner of strategists, never once failing to win in many contests.

After that I went from province to province duelling with strategists of various schools, and not once failed to win even though I had as many as sixty encounters. This was between the ages of thirteen and twenty-eight or twenty-nine.

When I reached thirty I looked back on my past. The previous victories were not due to my having mastered strategy. Perhaps it was natural ability, or the order of heaven, or that other schools' strategy was inferior. After that I studied morning and evening searching for the principle, and came to realise the Way of strategy when I was fifty.

Since then I have lived without following any particular Way. Thus with the virtue of strategy I practice many arts and abilities - all things with no teacher. To write this book I did not use the law of Buddha or the teachings of Confucius, neither old war chronicles nor books on martial tactics. I take up my brush to explain the true spirit of this Ichi school as it is mirrored in the Way of heaven and Kwannon. The time is the night of the tenth day of the tenth month, at the hour of the tiger (3-5 a.m.)

The Earth Scroll

Strategy is the craft of the warrior. Above all someone in a commander's position must practice it, and even troopers should know this Way. Yet today there is no warrior in the world who really understands the Way of strategy. (The term hyoho, here rendered as strategy, has a variety of meanings in Japanese, from large-scale military strategy, to martial arts or sword-fencing, and even - through play of words - the arts of peace and proper government. 'Strategy' is used as the translation largely for convenience, this should not be taken to imply that the other meanings are less appropriate - Musashi may even be referring to all of them at once.)

First let us illustrate the idea of a Way. There are various Ways. There is the Way of salvation by the law of Buddha, the Way of Kon Fu Tzu governing learning, the Way of healing as a physician, as a poet teaching the

Way of Waka, (a type of poem). Others follow the way of tea, archery, and many arts and skills. Each man practices as he feels inclined. Few are those that prefer the Way of martial strategy.

First, as is often said, a samurai must have both literary and martial skills: to be versed in the two is his duty. Even if he has no natural ability, a samurai must train assiduously in both skills to a degree appropriate to his status. On the whole, if you are to assess the samurai's mind, you may think it is simply attentiveness to the manner of dying. When it comes to the manner of dying, of course, there is no difference between the samurai, priests, women, and even peasants; everyone must know his obligations, think of what would be disgraceful, and be prepared for death when the moment comes. The samurai pursues martial strategy, however, in order to excel in everything, be it winning a duel or winning a combat with several men, be it for the benefit of your master or to establish your own reputation and distinguish yourself. The samurai does these things through the virtue of strategy. Some people may think that even if you learn martial arts, they will be useless in actual battles. That may be so, but the true spirit of martial strategy requires that you train to be useful at any moment and teach men so that they may become useful in everything.

The Way of Strategy (Hyoho no michi)

In China and Japan, practitioners of the Way have been known as "masters of strategy". Warriors should all learn this Way.

Recently people making a living as strategists are usually just teachers of sword techniques. The attendants of the Kashima and Katori shrines of the Hitachi province claim they received instruction from the Gods, and established schools based on this teaching, traveling from province to province to pass on their instruction. But this is only a recent meaning of the term strategy. (The Kashima shrine is dedicated to the warrior deity Takemikazuchi no Mikoto, the Katori shrine to his colleague Futsunushi no Kami. The Tenshin Shoden Katori Shinto ryu is documented to have existed in Musashi's time.)

Since ancient times Strategy (Hyoho) has been included among the juno (ten skills) and hachigei (eight arts) as rikata (profitable measures - divine favour in Buddhist Law, in other words the way benefiting oneself and others). Truly, rikata is one of the arts, although it is not just limited to standard sword techniques. The true value of swordsmanship cannot be seen solely by means of sword techniques. Needless to say swordsmanship limited to techniques alone can never rival the principles of Strategy.

Today we see the arts for sale. Men sell their own selves as commodities. As with the nut and the flower, the nut has become less important than the flower. In this kind of strategy, both those teaching and those learning the way are concerned with flamboyant style and showing off their technique, trying to hasten the bloom of the flower with commercial popularization. They speak of "this Dojo" and "that Dojo". They are looking only for quick benefits. Someone once said "Amateuristic strategy is the cause of serious grief". That was a true saying.

Generally speaking, there are four Ways in which men pass through life: as warriors, farmers, artisans and merchants.

The Way of the farmer is in using agricultural instruments; he sees springs through to autumns with an eye on the changes of season.

Second is the Way of the merchant. The Sake brewer obtains his ingredients and puts them to use, making his living from the profit he gains according to the quality of the product. Whatever the business, the merchant exists only by taking profit. This is the Way of the merchant.

Third is the gentleman warrior, carrying the weaponry of his Way. The warrior has to master the various properties and virtues of his different weapons. If a gentleman dislikes martial arts he will not appreciate the specific advantages of each weapon. For a member of a warrior house this shows a lack of culture.

Fourth is the Way of the artisan. The Way of the carpenter (architect and builder, all buildings were of wood) is to become proficient in the construction and use of his tools, to lay his plans correctly using the square and ruler, and then perform his work diligently according to the plan. Thus he passes through life.

These are the four Ways of life, of the warrior, the farmer, the artisan and the merchant.

I will now illustrate the way of strategy by likening it to the way of the craftsman.

The comparison with carpentry is a metaphor in reference to the notion of houses. We speak of houses of the nobility, houses of warriors, the Four houses (there are also four different schools of tea), ruin of houses, thriving of houses, the style of the house, the tradition of the house, and the name of the house. Since we refer to houses all the time, I have chosen the carpenter as a metaphor.

The word for carpenter is written as "great skill" or "master plan", and the Way of strategy is similar in that it requires great skill and masterful planning.

If you want to learn the art of strategy, ponder over this book. Let the teacher be as a needle, the student as a thread. You must practice constantly.

Comparing the Way of the carpenter to strategy

(In the following section, Musashi demonstrates his detailed knowledge of carpentry in comparing a command structure to the carpenters' guild. However, comparing samurai to carpenters appears to have been common in his days.)

The master carpenter is the organizer and director of the carpenters and it is his duty to understand the regulations of the country and the locality, and to abide by the rules of his guild.

The master carpenter must know the architectural theory of towers and temples, the plans of palaces and all sorts of structures, and must employ people to raise up houses. In this way, the Way of the master carpenter is comparable to the Way of the commander of a warrior house.

In the construction of houses, careful selection of woods is made. Straight unknotted timber of good appearance is used for the revealed pillars, straight timber with small defects is used for the inner pillars. Timbers of the finest appearance, even if a little weak, is used for the thresholds, lintels, doors, and sliding screens. Good strong timber, though it be gnarled and knotted, can always be used thoughtfully in consideration of the strengths of the other members of the house. Then the house will last a long time.

Even timber which is weak or knotted and crooked can be used as scaffolding, and later for firewood.

The master carpenter distributes the work among his men according to their levels of skill. Some are floor layers, others makers of sliding doors, thresholds and lintels, ceilings and so on. Those of poor ability lay the floor joists, and those of even lesser ability can carve wedges and do such miscellaneous work. If the master knows and deploys his men well the work will progress smoothly and the result will be good.

The master carpenter should take into account the abilities and limitations of his men. Circulating among them, he can know their spirit and different levels of morale, encourage them when necessary, understand what can and cannot be realized, and thus ask nothing unreasonable. The principle of strategy is like this.

The Way of Strategy

Like a soldier, the carpenter sharpens his own tools. He carries his equipment in his tool box, and works under the direction of his foreman. He makes columns and girders with an axe, shapes floorboards and shelves with a plane, cuts fine openwork and bas reliefs accurately, giving as excellent a finish as his skill will allow. This is the craft of the carpenters. When the carpenter becomes skilled, he works efficiently and according to correct measures. When he has developed practical knowledge of all the skills of the craft, he can become a master carpenter himself.

The carpenter will make it a habit of maintaining his tools sharp so they will cut well. Using these sharp tools masterfully, he can make miniature shrines, writing shelves, tables, paper lanterns, chopping boards and pot-lids. These are the specialties of the carpenter. Things are similar for the soldier. You ought to think deeply about this.

The attainment of the carpenter is that his work is not warped, that the joints are not misaligned, and that the work is truly planed so that it meets well together and is not merely finished in disjoint sections. This is essential.

If you want to learn this Way, deeply consider all the things written in this book one at a time. You must do sufficient research.

Outline of the Five Volumes of this Book of Strategy

In order to explain different aspects of the Way in individual sections, I have written this book in five volumes. These are entitled Earth, Water, Fire, Wind, and Void. (the void, or Nothingness, is a Buddhist term for the illusionary nature of worldly things.)

In the Earth book, I give an overall picture of the art of fighting and my own approach. It is difficult to know the true Way through swordsmanship alone. From large places one knows small places, from the shallows one goes to the depths. Because a straight road is made by leveling the earth and hardening it with gravel, I call the first volume Earth, as if it were a straight road mapped out on the ground.

Second, the Water volume. We make water our model and turn our spirit into water. Water adjusts itself to a square or round vessel with ease, turns itself into a single droplet or into a vast ocean. (Possibly a reference to Lao Tzu: Supreme good is like water. Water benefits everything, yet doesn't compete). It has the color of aquamarine depths. With that clarity I will write out my approach in this volume.

Once you definitely understand the principle of swordsmanship, the ability to defeat a single person at will means the ability to defeat all the people of the world. The spirit of defeating one man is the same for defeating ten million men. A commander's strategy, which requires him to make something large out of something small, is comparable to the making of a giant Buddhist statue from a foot-high scale model. I cannot write in detail how this is done. The principle of strategy is to know ten thousand things by having one thing. A few things of My Niten Ichi School are explained in this Water book.

Third is the Fire volume. This book is about fighting. The spirit of fire is fierce, whether the fire be small or big; and so it is with battles. The Way to do battle is the same for man to man fights and for the clash of armies of ten thousands. You must appreciate that the spirit can become big or small. Big things are easy to see: small things are difficult to see. In other words, it is difficult for large numbers of men to change position, so their movements can be easily predicted.

An individual can change his mind so quickly that his movements are difficult to predict. You must appreciate this. You must train day and night in order to get used to the quick changes of the mind in battle, until you begin to perceive them naturally. It is necessary to treat your training as part of everyday life. Then your spirit will remain unchanging.

Thus combat in battle is described in the Fire volume.

Fourthly the Wind volume (also the schools volume). This book is not concerned with my own Niten Ichi school but with other schools of strategy. By the term Wind I mean style or tradition, as in old traditions, present-day traditions, and family traditions of strategy. Here I clearly explain the strategies of the various schools in the world. This is tradition. It is difficult to know yourself if you do not know others (a possible reference to Sun Tzu: If you know your enemy and yourself, you will not be in danger in a hundred battles). To all Ways there are side-tracks. Even if you study your Way daily, if your spirit diverges, objectively it is not the true Way, even though you may think you are following a good Way. What may be a little divergence in the beginning will later become a large divergence. You must realize this.

Strategy has come to be thought of as mere sword-fencing, and it is not unreasonable that this should be so. The benefit of my strategy, although it includes sword-fencing, lies in a different matter entirely. I have explained what is commonly meant by strategy in other schools in the Wind (Tradition) book.

Fifthly, the book of the Void. By void I mean that which has no beginning and no end. What can you call its innermost depths or its entrance? Attaining this principle means letting go of principles. The Way of strategy is the Way of nature. It has its own freedom. Once you detach from principles, you will acquire exceptional skill spontaneously and independently. Once you know the power of nature, you will discern the rhythm of any situation, and you will hit the enemy naturally with your every strike. All this is the Way of the Void. I intend to show how to enter the true Way naturally in the book of the Void.

Two Swords as One

Samurai, both commanders and troopers, carry two swords at their belt. In olden times these were called the tachi (long sword) and the katana (sword); nowadays they are known as the katana (sword) and the wakizashi (companion sword). I don't need to explain this in detail. Let it suffice to say that in Japan, a warrior carries two swords as a matter of duty, whether he knows how to use them or not. It is the Way of the warrior.

I have decided to call my approach "Nito Ichi Ryu" (two swords as one) to show the advantages of using both swords.

As for the spear and the halberd, and so on, these are considered additional weapons.

Students of the Nito Ichi School of strategy should train from the start with the sword and the long sword in either hand. This is a truth: when you sacrifice your life, you would want to make fullest use of your available weaponry. It is unnatural not to do so, and to die with a weapon yet undrawn.

If you hold a sword with both hands, it is difficult to wield it freely to the left and to the right, so my method is to get used to wielding the sword in one hand. This does not apply to large weapons such as the spear or halberd, but swords and companion swords can be used in one hand. It is encumbering to hold a sword in both hands when you are on horseback, when running on uneven roads, on swampy ground, muddy rice fields, stony ground, or in a crowd of people. To hold the long sword in both hands is not the true Way, for if you carry a bow or spear or other arms in your left hand you have only one hand free for the long sword. If or when you find it difficult to cut an enemy down with one hand, then by all means use both hands. There is nothing complicated about this.

It is not difficult to wield a sword in one hand; the Way to learn this is to train with two long swords, one in each hand. It will seem difficult at first, but everything is difficult at first. Bows are difficult to draw, halberds are difficult to wield. In each case, you get used to the tool: as you become accustomed to the bow your pull will become stronger, and as you become used to wielding the long sword, you will gain the power of the Way and wield the sword easily.

As I will explain in the second book, the Water Book, there is no fast way of wielding the long sword. The long sword should be wielded broadly and the companion sword closely. This is the basic idea for the beginner.

In my school, you can win with a long weapon, and yet you can also win with a short weapon. For this reason I don't specify the length of the sword but regard the essence of my approach as the resolve to gain victory by any means, whatever the weapon and whatever its size.

It is better to use two swords rather than one when you are fighting a mob, and especially if you want to take a prisoner.

These things need not be explained in detail. From each point, ten thousand things can be inferred. When you attain the Way of strategy there will be nothing you cannot see. You must study hard.

The Principles behind the Characters reading "Strategy"

Masters of the long sword are traditionally known as heihosha (strategists). As for the other military arts, those who master the bow are called archers, those who master the spear are called spearmen, those who master the gun are called marksmen, and those who master the halberd are called halberdiers. But we do not call masters of the long sword "long swordsmen", nor do we speak of "short swordsmen". Bows, guns, spears and halberds are all tools of the warriors and each should be a way to master strategy.

Nevertheless, the sword alone is associated with mastery of strategy. There is a reason for this. To master the virtue of the long sword is to govern the world and oneself, thus the long sword is the basis of strategy (this observation is based on ancient Japanese sword worship). The principle is "strategy by means of the long sword". If he attains the virtue of the long sword, one man can beat ten men. Just as one man can beat ten, so a hundred men can beat a thousand, and a thousand can beat ten thousand. In my strategy, things are no different for one man or for ten thousand. This strategy is the complete warrior's craft.

The Way of the warrior does not include other Ways, such as Confucianism, Buddhism, tea, artistic accomplishments and dancing. But even though these are not part of your Way, if you know the Way broadly you will see it in everything. It is essential for each of us as human beings to polish our individual Way.

The Advantages of Weapons in Strategy

There is an appropriate time and place for the use of the various weapons.

The companion sword is most useful in a confined space, or when you are engaged at close quarters with an opponent. The long sword can generally be used effectively in any situation.

The halberd is inferior to the spear on the battlefield. With the spear you take the initiative, the halberd is more defensive. In the hands of men of equal ability, the spear gives a little extra strength. Spear and halberd both have their uses depending on the situation, but neither is very helpful in a confined space. They are also not appropriate for taking a prisoner. They are essentially weapons for the field.

Anyway, if you only learn to use weapons indoors, you will narrow your focus to unimportant details and forget the true way. Thus you will have difficulty in actual encounters.

The bow is tactically strong in both charges and retreats, especially during battles on an open field, as it is possible to shoot quickly from among the ranks of the spearmen or others. However, it is inadequate in sieges, or when the enemy is more than forty yards away. Nowadays the schools of archery as well as other arts have more flowers than fruit. In times of real need such kind of skill is useless.

From inside fortifications, the gun has no equal among weapons. It is the supreme weapon on the field before the ranks clash, but once swords are crossed the gun becomes inadequate.

One of the virtues of the bow is that you can see the arrows in flight and correct your aim accordingly, whereas the path of a bullet cannot be seen. You must appreciate the importance of this.

Your horse should have good endurance and no defects. To summarize the essentials on weapons, horses should walk strongly, and swords and short swords should cut strongly. Spears and halberds must penetrate strongly and stand up to heavy use, bows and guns must be strong and accurate. Weapons should be sturdy rather than decorative.

You should not have a favorite weapon, or any other exaggerated preference for that matter. To become overly attached to one weapon is as bad as not knowing it sufficiently well. You should not imitate others, but use those weapons which suit you, and which you can handle properly. It is bad for both commanders and troopers to entertain likes and dislikes. Pragmatic thinking is essential. These are things you must learn thoroughly.

Rhythm in strategy

There is rhythm in everything, however, the rhythm in strategy, in particular, cannot be mastered without a great deal of hard practice.

Among the rhythms readily noticeable in our lives are the exquisite rhythms in dancing and accomplished pipe or string playing. Timing and rhythm are also involved in the military arts, shooting bows and guns, and riding horses. In all skills and abilities there is timing.

There is also rhythm in the Void.

There is a rhythm in the whole life of the warrior, in his thriving and declining, in his harmony and discord. Similarly, there is a rhythm in the Way of the merchant, of becoming rich and of losing one's fortune, in the rise and fall of capital. All things entail rising and falling rhythm. You must be able to discern this. In strategy there are various considerations. From the outset you must attune to your opponent, then you must learn to disconcert him. It is crucial to know the applicable rhythm and the inapplicable rhythm, and from among the large and small rhythms and the fast and slow rhythms find the relevant rhythm. You must see the rhythm of distance, and the rhythm of reversal. This is the main thing in strategy. It is especially important to understand the rhythms of reversal; otherwise your strategy will be unreliable.

In combat, you must learn the rhythm of each opponent, and use the rhythms that your opponents don't expect. You win by creating formless rhythms out of the rhythm of the Void.

All the five books are chiefly concerned with rhythm. You must train sufficiently to appreciate this.

If you practice diligently day and night in the strategy outlined above, your spirit will ally broaden. Thus you will come to prehend large scale strategy and the ategy of one on one combat. This is recorded for the first time in these five volumes of Earth, Water, Fire, Wind, and Void.

Those who sincerely desire to learn my way of strategy will follow these rules for learning the art:

1. Do not harbor sinister designs. Think honestly and truthfully.
2. The Way is in training. One must continue to train.
3. Cultivate a wide range of interests in the ten skills and ten arts. Then one can definitely find the benefits of hyoho and develop oneself.
4. Be knowledgeable in a variety of occupations, and learn the thinking of people who work in them.
5. Know the difference between loss and gain in worldly matters.
6. Nurture the ability to perceive the truth in all matters. It is important to build up an intuitive judgment and understand true values.

7. Be aware of those things which cannot be easily seen with the eye. Develop intuitive judgment and a mind that freely controls one's body.

8. Do not be negligent, but pay attention even to the smallest details. Keep them in mind all the time, so as to avoid unexpected failure.

9. Do not engage in useless activity. Do not argue about useless things. Concentrate on your duties.

It is important to start by setting these broad principles in your heart, and train in the Way of strategy. If you do not look at things in a larger context it will be difficult for you to master strategy. If you learn and attain these principles, you will never lose even in individual combat against twenty or thirty enemies. First of all, you must set your heart on strategy and work earnestly while sticking to the Way. With time you will be able to beat men with your hands, and to defeat people by using your eyes. When through training you become able to freely control your own body, you can conquer men with your body. And because strategy develops the mind, with sufficient training you will be able to beat people with your spirit. When you have reached this point, will it not mean that you are invincible?

Moreover, in large scale strategy the superior man will attract and keep able subordinates, bear himself correctly, govern the country and care for the people, thus preserving the ruler's discipline.

The Way of strategy is to be self-reliant, not losing at anything, to guide others, to gain benefits and honor, and to make peace with others.

The Water Scroll

The spirit of the Ni Ten Ichi school of strategy is based on Water, and this Water Book explains methods of victory as the long-sword form of the Niten Ichi school. Language does not suffice to explain the Way in detail, but it can be grasped intuitively. Study this book; read a word then ponder on it. If you interpret the meaning loosely you will mistake the Way.

The principles of strategy are written down here in terms of single combat, but you must think broadly, so that you attain an understanding for ten-thousand-a-side battles.

Strategy is different from other things in that if you mistake the Way even a little you will become bewildered and fall into bad ways.

You will not reach the Way of strategy by merely reading this book. Absorb the things written in this book. Do not just read, memorize or imitate, but make sure that you realize the principle from within your own heart. Study hard to absorb these things into your body.

Spiritual Bearing in Strategy

In strategy your spiritual bearing must not be any different from normal. In normal times, and in times of combat, try to be no different: Keep your mind broad and straight; do not stretch it taut; do not allow it to grow in the least lax; do not make it lean to one side but hold it at the center; keep it quietly fluid, doing your best to maintain it in a fluid state even while it is fluid.

When you are quiet, your mind shouldn't be quiet; when you are moving fast, your mind shouldn't at all be moving fast. Even when your mind is calm do not let your body relax, and when your body is relaxed do not let your mind slacken. Do not let your spirit be influenced by your body, nor your body be influenced by your spirit. Your mind should lack nothing while having no excess. An elevated spirit is weak and a low spirit is weak. Superficially you may have your mind appear weak, but you must keep it strong inwardly, lest people can tell what you really are. Do not let the enemy see your spirit.

If you have a small body, you must know whatever there is to know about having a large body; if you have a large body, you must know all about having a small body. Whatever your size, you must keep your mind straight and not be misled by knowing only your own body. With your spirit open and unconstricted, look at things from a high point of view. You must cultivate your wisdom and spirit. Polish your wisdom: learn public justice, learn to distinguish between right and wrong, study the Ways of different arts one by one to experience what is sought in each. When you cannot be deceived by anyone you will have acquired judgment in strategy.

The wisdom of strategy is different from other things. On the battlefield, even when you are hard-pressed, you should ceaselessly research the principles of strategy so that you can develop a steady spirit.

Physical Bearing in Strategy

In holding your body, your face shouldn't be downcast or upturned, tilted or twisted.

Do not allow your eyes to be distracted easily. Do not knit your brow, but keep the space between your eyebrows wrinkled, lest your eyes roll. Taking care not to blink, narrow your eyes a little. With your face relaxed, keep your nose straight with a feeling of slightly flaring your nostrils, your lower jaw a little forward. As for your head, keep the muscles in back straight, your nape tight. Treat your body from shoulders down as one. Hold both shoulders down, your spine erect. Do not stick your buttocks out. Put strength into your legs from knees to toes. Thrust your belly out lest you bend at the hips. There is something called "wedging-in": you put the weight of your belly on the scabbard of your short sword lest your belt slacken.

On the whole, in strategy it is most important that you regard your normal bearing as the same as your bearing at a time of fighting, and your bearing at a time of fighting as the same as your normal bearing. You must research this well.

The Gaze in Strategy

You eye things in a sweeping, broad fashion. As for the two manners of seeing things, kan (observing) and ken (seeing), the eye for kan is strong, the eye for ken weak; seeing distant things as if they are close at hand and seeing close things as if they are distant is special to the art of fighting. Knowing your opponent's sword and yet not in the least seeing it (not being distracted by insignificant movements) is important in the art of fighting. You must study this. The gaze is the same for single combat and for large-scale strategy.

It is necessary in strategy to be able to look to both sides without moving the eyeballs. You cannot master this ability quickly. Learn what is written here; use this gaze in everyday life and do not vary it whatever happens.

Holding the Long Sword

Grip the long sword lightly with your thumb and forefinger, with the middle finger neither tight nor slack, and with the last two fingers tight. It is bad to leave slack in your hands.

When you take up a sword, you must do it with the intent of cutting the enemy. As you cut an enemy you must not change your grip, and your hands must not flinch. When you dash the enemy's sword aside, or ward it off, or force it down, you may only change your thumb and forefinger a little. Above all, you must grip the sword with the intent of cutting the enemy.

The grip for combat and for sword-testing is the same, you always grip the sword as if you want to kill a man.

Generally speaking, stiffness is to be avoided, in both sword and hands. Stiffness leads to death. A living hand is flexible. (Another possible reference to Lao Tzu) You must bear this in mind.

Footwork

With the tips of your toes somewhat floating, tread firmly with your heels. Whether you move fast or slow, with large or small steps, your feet should always move naturally as in normal walking. Avoid jumping steps, floating steps and stomping.

An important concept in my school is called complementary ("Yin-Yang") stepping: this means that you do not move one foot alone. You should always move your feet in complementary steps, left-right and right-left when cutting, withdrawing, or warding off a cut. You should not move on one foot alone.

The Five Attitudes

The five attitudes are: Upper, Middle, Lower, Right Side, and Left Side. Although attitude has these five divisions, the one purpose of all of them is to cut the enemy. There are none but these five attitudes.

(Kamae (attitude), from the verb kamaeru: to build, set up, adopt a stance, posture or defensive attitude)

Whatever attitude you are in, do not be conscious of adopting the attitude; think only of cutting.

Your attitude should be large or small according to the situation. Upper, Lower and Middle attitudes are decisive. Left Side and Right Side attitudes are fluid. Left and Right attitudes should be used if there is an obstruction overhead or to one side. The decision to use Left or Right depends on the place.

To understand the essence of the Way, you must thoroughly understand the middle attitude. The middle attitude is the heart of attitudes. If we look at strategy on a broad scale, the middle attitude is the seat of the commander, with the other four attitudes following the commander. You must appreciate this.

The Way of the Long Sword

If we know the Way of the long sword well, we can easily wield the sword we usually carry, even with two fingers.

If you try to wield the long sword unnaturally fast, you are mistaken. To wield the long sword well you must wield it calmly. If you try to wield it quickly, like a folding fan or a short sword, you will deviate from the Way by using what is called "knife whittling". The long sword is hard to wield this way, and you cannot cut down a man efficiently with a long sword in this manner.

When you have cut downwards with the long sword, lift it straight back up along a natural path; when you have cut sideways, return the sword naturally along a sideways path. Always return the sword in a reasonable way. Extend the elbows broadly in a comfortable way, and wield the sword powerfully. This is the Way of the long sword.

If you learn to use the five approaches of my strategy, you will be able to wield a sword well. You must train constantly.

The Five Approaches

1. The first approach is the Middle attitude. Confront the enemy with the point of your sword against his face. When he attacks, dash his sword to the right and "ride" it. Or, when the enemy attacks, deflect the point of his sword by hitting downwards, keep your long sword where it is, and as the enemy renews his attack cut his arms from below. This is the first method.

The five approaches are this kind of thing. You must train repeatedly using a long sword in order to learn them. When you master my Way of the long sword, you will be able to control any attack the enemy makes. I assure you, there are no attitudes other than the five attitudes of the long sword of Ni To.

2. In the second approach with the long sword, from the Upper attitude cut the enemy just as he attacks. If the enemy evades the cut, keep your sword where it is and, scooping up from below, cut him as he

renews the attack. It is possible to repeat the cut from here.

In this method there are various changes in timing and spirit. You will be able to understand this by training in the Niten Ichi school. You will always win with the five long sword methods. You must train repetitively.

3. In the third approach, adopt the Lower attitude, anticipating scooping up. When the enemy attacks, hit his hands from below. As you do so he may try to hit your sword down. If this is the case, cut his upper arm(s) horizontally with a feeling of "crossing". This means that from the lower attitudes you hit the enemy at the instant that he attacks.

You will encounter this method often, both as a beginner and in later strategy. You must train holding a long sword.

4. In this fourth approach, adopt the Left Side attitude. As the enemy attacks hit his hands from below. If as you hit his hands he attempts to dash down your sword, with the feeling of hitting his hands, parry the path of his long sword and cut across from above your shoulder.

This is the Way of the long sword. Through this method you win by parrying the line of the enemy's attack. You must research this.

5. In the fifth approach, the sword is in the Right Side attitude. In accordance with the enemy's attack, cross your long sword from below at the side to the Upper attitude. Then cut straight from above.

This method is essential for knowing the Way of the long sword well. If you can use this method, you can freely wield a heavy long sword.

I cannot describe in detail how to use these five approaches. You must become well acquainted with my "in harmony with the long sword" Way, learn large-scale timing, understand the enemy's long sword, and become used to the five approaches from the outset. You will always win by using these five methods, with various timing considerations discerning the enemy's spirit. You must consider all this carefully.

Attitude No-Attitude

"Attitude No-Attitude" means that you should not intentionally take specific long sword attitudes. Though the attitudes are differentiated into five types, they are all meant only to cut the enemy.

Even though you can't help holding your sword in one of the five ways of holding the long sword, you must hold the sword in such a way that it is easy to cut the enemy well, in accordance with the situation, the place, and the move of the enemy. You may start from the Upper attitude, but if you lower your sword a bit you adopt the Middle attitude, and from the Middle attitude you can raise the sword a little in your technique and adopt the Upper attitude. From the lower attitude you can raise the sword and adopt the Middle attitudes as the occasion demands. According to the situation, if you turn your sword from either the Left Side or Right Side attitude towards the center, the Middle or the Lower attitude results.

This is why I say you are taking an attitude without taking an attitude. The principle of this is called "Existing Attitude - Nonexisting Attitude".

In any event, once you take a sword in your hands, you must be prepared to cut apart the enemy, whatever the means. Whenever you parry, hit, catch, strike or block the enemy's attacking sword, you must know the opportunities to cut the enemy in the same movement. It is essential to attain this. If you think only of catching, blocking, striking or tying up the enemy, you will not be able to actually kill him. More than anything, you must be thinking of carrying your every movement through to the kill. You must thoroughly research this.

Attitude in strategy on a larger scale is called "Battle Array". Such attitudes are all aimed at winning the battle. Fixed formation is bad. Study this well.

To Hit the Enemy In One Timing

"In One Timing" means, when you have closed with the enemy, to hit him as quickly and directly as possible, without moving your body or settling your spirit, while you see that he is still undecided. The timing of hitting before the enemy decides to withdraw, break or hit, is this "In One Timing".

You must train to achieve this timing, to be able to hit in the timing of an instant.

The Body Timing of Two

When you attack and the enemy quickly retreats, as you see him tense you must feint a cut. Then, as he relaxes, follow up and hit him. This is the "Body Timing of Two".

It is very difficult to attain this by merely reading this book, but you will soon understand with a little instruction.

No Design, No Conception

(Munen musou, when word and actions are spontaneously the same, is the ultimate state of consciousness in Buddhism. Musou is equivalent to the Sanskrit animitta)

In this method, when the enemy attacks and you also decide to attack, your body and mind turn into a single striking movement and your hands strike out of the Void naturally, swiftly and strongly. This is the "No Design, No Conception" cut.

This is the most important method of hitting. It is often used. You must train hard to understand it.

The Flowing Water Cut

The "Flowing Water Cut" is used when you are struggling blade to blade with the enemy. When he breaks and quickly withdraws trying to spring with his long sword, expand your body and spirit and cut him as slowly as possible with your long sword, following your body like stagnant water. You can cut with certainty if you learn this. You must discern the enemy's grade.

Continuous Cut

When you attack and the enemy also attacks and your swords spring together, in one action cut his head, hands and legs. When you cut several places with one sweep of the long sword, it is the "Continuous Cut". You must practice this cut; it is often used. With detailed practice you should be able to understand it.

The Fire and Stones Cut

The Fires and Stones Cut means that when the enemy's long sword and your long sword clash together you cut as strongly as possible without raising the sword even a little. This means cutting quickly with the hands, body and legs - all three cutting strongly. If you train well enough you will be able to strike strongly.

The Red Leaves Cut

The Red Leaves Cut (allusion to falling, dying leaves) means knocking down the enemy's long sword. The spirit should be getting control of his sword. When the enemy is in a long sword attitude in front of you and intent on cutting, hitting and parrying, you strongly hit the enemy's long sword with the Fire and Stones Cut, perhaps in the spirit of the "No Design, No Conception" Cut. If you then beat down the point of his sword with a sticky feeling, he will necessarily drop the sword. If you practice this cut it becomes easy to make the enemy drop his sword. You must train repetitively.

The Body in Place of the Long Sword

Also "the long sword in place of the body". Usually we move the body and the sword at the same time to cut the enemy. However, according to the enemy's cutting method, you can dash against him with your body first, and afterwards cut with the sword. If his body is immovable, you can cut first with the long sword, but generally you hit first with the body and then cut with the long sword. You must research this well and practice hitting with your body.

Cut and Slash

To cut and to slash are two different things. Cutting, whatever form of cutting it is, is decisive, with a resolute spirit. Slashing is nothing more than touching the enemy. Even if you slash strongly, and even if the enemy dies instantly, it is still slashing. When you cut, your spirit is resolved. You must appreciate this. If you first slash the enemy's hands or legs, you must then cut strongly. Slashing is in spirit the same as touching.

When you realize this, they become indistinguishable. Learn this well.

Chinese Monkey's Body

The Chinese Monkey's Body (short-armed monkey) is the spirit of not stretching out your arms. The spirit is to get in quickly, without in the least extending your arms, before the enemy cuts. If you are intent upon not stretching out your arms you are effectively far away, the spirit is to go in with your whole body. When you come to within arm's reach it becomes easy to move your body in. You must research this well.

Glue and Lacquer Emulsion Body

The spirit of "Glue and Lacquer Emulsion Body" is to stick to the enemy and not separate from him. When you approach the enemy, stick firmly with your head, body and legs. People tend to advance their head and legs quickly, but their body lags behind. You should stick firmly so that there is not the slightest gap between the enemy's body and your body. You must consider this carefully.

To Strive for Height

By "to strive for height" is meant, when you close with the enemy, to strive with him for superior height without cringing. Stretch your legs, stretch your hips, and stretch your neck face to face with him. When you think you have won, and you are the higher, thrust in strongly. You must learn this.

To Apply Stickiness

When the enemy attacks and you also attack with the long sword, you should go in with a sticky feeling and fix your long sword against the enemy's as you receive his cut. The spirit of stickiness is not hitting very strongly, but hitting so that the long swords do not separate easily. It is best to approach as calmly as possible when hitting the enemy's long sword with stickiness. The difference between "Stickiness" and "Entanglement" is that stickiness is firm and entanglement is weak. You must appreciate this.

The Body Strike

The Body Strike means to approach the enemy through a gap in his guard. The spirit is to strike him with your body. Turn your face a little aside and strike the enemy's breast with your left shoulder thrust out. Approach with the spirit of bouncing the enemy away, striking as strongly as possible in time with your breathing. If you achieve this method of closing with the enemy, you will be able to knock him ten or twenty feet away. It is possible to strike the enemy until he is dead. Train well.

Three Ways to Parry His Attack

There are three methods to parry a cut:

First, by dashing the enemy's long sword to your right, as if thrusting at his eyes, when he makes an attack.

Or, to parry by thrusting the enemy's long sword towards his right eye with the feeling of snipping his neck.

Or, when you have a short "long sword", without worrying about parrying the enemy's long sword, to close with him quickly, thrusting at his face with your left hand.

These are the three methods of parrying. You must bear in mind that you can always clench your left hand and thrust at the enemy's face with your fist. For this it is necessary to train well.

To Stab at the Face

To stab at the face means, when you are in confrontation with the enemy, that your spirit is intent of stabbing at his face, following the line of the blades with the point of your long sword. If you are intent on stabbing at his face, his face and body will become readable. When the enemy becomes as if readable, there are various opportunities for winning. You must concentrate on this. When fighting and the enemy's body becomes as if readable, you can win quickly, so you ought not to forget to stab at the face. You must pursue the value of this technique through training.

To Stab at the Heart

To stab at the heart means, when fighting and there are obstructions above, or to the sides, and whenever it is difficult to cut, to thrust at the enemy. You must stab the enemy's breast without letting the point of your long sword waver, showing the enemy the ridge of the blade square-on, and with the spirit of deflecting his long sword. The spirit of this principle is often useful when we become tired or for some reason our long sword will not cut. You must understand the application of this method.

To Scold

"Scold" means that, when the enemy tries to counter-cut as you attack; you counter-cut again, coming up from below as if thrusting at him. Both strikes should follow in rapid succession, scolding the enemy. Thrust upwards, "Tut!", then cut "TUT!" This move can be used time and time again in a duel. The way to scold is to raise your sword as if to thrust the enemy, then to slash simultaneously. You should study this rhythm through repetitive practice.

The Smacking Parry

By "smacking parry" is meant that when you clash swords with the enemy, you meet his attacking cut on your long sword with a tee-dum, tee-dum rhythm, smacking his sword and cutting him. The spirit of the smacking parry is not parrying, or smacking strongly, but smacking the enemy's long sword in accordance with his attacking cut, primarily intent on quickly cutting him. If you understand the timing of smacking, however hard your long swords clash together, your sword point will not be knocked back even a little. You must research sufficiently to realize this.

Facing Many Enemies

Facing Many Enemies applies when you must fight alone against many enemies. Draw both the long and short swords, and spread them wide to the left and right in a horizontal attitude.

Even though they come from all four directions, your aim is to chase the enemies to one place. Observe their attacking order, and advance quickly to meet him who attack first. Sweep your eyes around broadly to remain aware of the overall situation, carefully examining the stances of the enemies and their attacking order, and cut left and right simultaneously in different directions, swinging both swords without mutual interference. It is not good to pause after cutting to different directions. You must quickly re-assume your original attitudes to both sides. Cut the next enemies down as they advance, crushing them in the direction from which they attack. Do your best to drive the enemy together, as if tying a line of fishes, and when they are seen to be piled up and entangled, cut them down strongly without giving them room to move.

If you frontally attack a crowd, you can hardly make progress. On the other hand, if you intend to cut one at the time each one that will advance first, you will be in a waiting attitude and not make progress. Respond to your enemies' rhythm, know their weakness and take advantage of it. Then you win.

If you practice with your friends often until you learn to force the whole group into a single file, you can deal as easily with one enemy as with ten or twenty. It requires thorough practice and examination.

The Advantage in Dueling

You can learn how to win a duel with the long sword through strategy, but it cannot be clearly explained in writing. You must practice diligently in order to understand how to win.

The true Way of strategy is revealed in the long sword. This is transmitted orally.

One Cut

You can win with certainty with the spirit of a single cut. It is difficult to attain this if you do not learn strategy well. If you practice this well, strategy will come from your heart and you will be able to win at will. You must train diligently.

Direct Penetration

The spirit of Direct Penetration is handed down in the true Way of the Ni To Ichi School.

Teach your body strategy. You must practice well. This is transmitted orally.

Epilogue

Recorded in the above book is an outline of my school of sword-fighting.

To learn how to win with the long sword in strategy, first learn the five approaches and the five attitudes, then absorb the Way of the long sword naturally in your body. You must sharpen the spirit to understand rhythm, handle the long sword naturally, and move your body with total freedom in harmony with your spirit. Whether beating one man or two, you will then know what is good or bad in strategy.

Study the contents of this book, practice one item at a time, and through fighting with enemies you will gradually come to know the principle of the Way of Strategy.

Deliberately, with a patient spirit, absorb the virtue of all this, from time to time raising your hand in combat. Maintain this spirit whenever you cross swords with and enemy.

Even a thousand-mile road is walked one step at a time. It is the duty of a warrior to study this art without hurry and practice it over the years. Try to defeat today what you were yesterday, defeat lesser men tomorrow, and stronger men the day after.

Train according to this book, not allowing your heart to be swayed along a side-track. Even if you defeat an enemy, if you do so in a way contrary to what you have learned, you are not following the true Way. If you grasp this principle, you will be able to defeat fifty or sixty men single-handedly. When that happens, you will have reached enlightenment in the way of strategy through swordsmanship, for large battles as well as individual combat. I call practice for a thousand days *tan* (hardening), practice for ten thousand days *ren* (practice. Musashi here divides the word *tanren* or drill. One thousand days refers to three years and ten thousand to thirty. Musashi's intention is explaining that one must continue to seek the way).

The Fire Scroll

In this Fire Book of the Ni To Ichi school of strategy I describe fighting as fire.

In the first place, people think narrowly about the benefit of strategy. By using only their fingertips, they only know the benefit of three of the five inches of the wrist. They let a contest be decided, as with the folding fan, merely by the span of their forearms. They specialize in the small matter of dexterity, learning such trifles as hand and leg movements with the bamboo practice sword.

In my strategy, the training for killing enemies is by way of many contests, fighting for survival, discovering the meaning of life and death, learning the Way of the sword, judging the strength of attacks and understanding the Way of the "edge and ridge" of the sword.

You cannot profit from small techniques particularly when full armor is worn. ("Roku

Gu" (six pieces): body amour, helmet, mask, thigh pieces, gauntlets and leg pieces.) My way of strategy is the sure method to win when fighting for your life one man against five or ten. There is nothing wrong with the principle "one man can beat ten, so a thousand men can beat ten thousand". You must research this. Of course you cannot assemble a thousand or ten thousand men for everyday training. But you can become a master of strategy by training alone with a sword, so that you can understand the enemy's stratagems, his strength and resources, and come to appreciate how to apply strategy to beat ten thousand enemies.

Any man who wants to master the essence of my strategy must research diligently, training morning and evening. Thus can he polish his skill, become free from self, and realize extraordinary ability. He will come to possess miraculous power.

This is the practical result of strategy.

Depending on the Place

Examine your environment.

Stand in the sun; that is, take up an attitude with the sun behind you. If the situation does not allow this, you must try to keep the sun on your right side. In buildings, you must stand with the entrance behind you or to your right. Make sure that your rear is unobstructed, and that there is free space on your left, your right side being occupied with your side attitude. At night, if the enemy can be seen, keep the fire behind you and the entrance to your right, and otherwise take up your attitude as above. You must look down on the enemy, and take up your attitude on slightly higher places. For example, the Kamiza (residence of the ancestral spirit of a house; often a slightly raised recess in a wall (with ornaments) in a house is thought of as a high place.

When the fight comes, always endeavor to chase the enemy around to your left side. Chase him towards awkward places, and try to keep him with his back to awkward places. When the enemy gets into an inconvenient position, do not let him look around, but conscientiously chase him around and pin him down. In houses, chase the enemy into the thresholds, lintels, doors, verandas, pillars, and so on, again not letting him see his situation.

Always chase the enemy into bad footholds, obstacles at the side, and so on, using the virtues of the place to establish predominant positions from which to fight. You must research and train diligently in this.

The Three Methods to Forestall the Enemy

The first is to forestall him by attacking. This is called Ken No Sen (to set him up).

Another method is to forestall him as he attacks. This is called Tai No Sen (to wait for the initiative).

The other method is when you and the enemy attack together. This is called Tai Tai No Sen (to accompany him and forestall him).

There are no methods of taking the lead other than these three. Because you can win quickly by taking the lead, it is one of the most important things in strategy. There are several things involved in taking the lead. You must make the best of the situation, see through the enemy's spirit so that you grasp his strategy and defeat him. It is impossible to write about this in detail.

The First - Ken No Sen

When you decide to attack, keep calm and dash in quickly, forestalling the enemy. Or you can advance seemingly strongly but with a reserved spirit, forestalling him with the reserve.

Alternatively, advance with as strong a spirit as possible, and when you reach the enemy move with your feet a little quicker than normal, unsettling him and overwhelming him sharply.

Or, with your spirit calm, attack with a feeling of constantly crushing the enemy, from first to last. The spirit is to win in the depths of the enemy.

These are all Ken No Sen.

The Second - Tai No Sen

When the enemy attacks, remain undisturbed but feign weakness. As the enemy reaches you, suddenly move away indicating that you intend to jump aside, then dash in attacking strongly as soon as you see the enemy relax. This is one way.

Or, as the enemy attacks, attack still more strongly, taking advantage of the resulting disorder in his timing to win.

This is the Tai No Sen Principle.

The Third - Tai Tai No Sen

When the enemy makes a quick attack, you must attack strongly and calmly, aim for his weak point as he draws near, and strongly defeat him.

Or, if the enemy attacks calmly, you must observe his movements and, with your body rather floating, join in with his movements as he draws near. Move quickly and cut him strongly.

This is Tai Tai No Sen.

These things cannot be clearly explained in words. You must research what is written here. In these three ways of forestalling, you must judge the situation. This does not mean that you always attack first; but if the enemy attacks first you can lead him around. In strategy, you have effectively won when you forestall the enemy, so you must train well to attain this.

To Hold Down a Pillow

"To Hold Down a Pillow" means not allowing the enemy's head to rise.

In contests of strategy it is bad to be led about by the enemy. You must always be able to lead the enemy about. Obviously the enemy will also be thinking of doing this, but he cannot forestall you if you do not allow him to come out. In strategy, you must stop the enemy as he attempts to cut; you must push down his thrust, and throw off his hold when he tries to grapple. This is the meaning of "to hold down a pillow". When you have grasped this principle, whatever the enemy tries to bring about in the fight you will see in advance and suppress it. The spirit is to check his attack at the syllable "at...", when he jumps check his jump at the syllable "ju...", and check his cut at "cu...".

The important thing in strategy is to suppress the enemy's useful actions but allow his useless actions. However, doing this alone is defensive. First, you must act according to the Way, suppressing the enemy's techniques, foiling his plans and thence command him directly. When you can do this you will be a master of strategy. You must train well and research "holding down a pillow".

Crossing at a Ford

"Crossing at a ford" means, for example, crossing the sea at a strait, or crossing over a hundred miles of broad sea at a crossing place. I believe this "crossing at a ford" occurs often in man's lifetime. It means setting sail even though your friends stay in harbor, knowing the route, knowing the soundness of your ship and the favor of the day. When all the conditions are met, and there is perhaps a favorable wind, or a tailwind, then set sail. If the wind changes within a few miles of your destination, you must row across the remaining distance without sail.

If you attain this spirit, it applies to everyday life. You must always think of crossing at a ford.

In strategy also it is important to "cross at a ford". Discern the enemy's capability and, knowing your own strong points, "cross the ford" at the advantageous place, as a good captain crosses a sea route. If you succeed in crossing at the best place, you may take your ease. To cross at a ford means to attack the enemy's weak point, and to put yourself in an advantageous position. This is how to win large-scale strategy. The spirit of crossing at a ford is necessary in both large- and small-scale strategy.

You must research this well.

To Know the Times

"To know the times" means to know the enemy's disposition in battle. Is it flourishing or waning? By observing the spirit of the enemy's men and getting the best position, you can work out the enemy's disposition and move your men accordingly. You can win through this principle of strategy, fighting from a position of advantage.

When in a duel, you must forestall the enemy and attack when you have first recognized his school of strategy, perceived his quality and his strong and weak points. Attack in an unsuspecting manner, knowing his metre and modulation and the appropriate timing.

Knowing the times means, if your ability is high, seeing right into things. If you are thoroughly conversant with strategy, you will recognize the enemy's intentions and thus have many opportunities to win. You must sufficiently study this.

To Tread Down the Sword

"To tread down the sword" is a principle often used in strategy. First, in large scale strategy, when the enemy first discharges bows and guns and then attacks it is difficult for us to attack if we are busy loading powder into our guns or notching our arrows. The spirit is to attack quickly while the enemy is still shooting with bows or guns. The spirit is to win by "treading down" as we receive the enemy's attack.

In single combat, we cannot get a decisive victory by cutting, with a "tee-dum tee-dum" feeling, in the wake of the enemy's attacking long sword. We must defeat him at the start of his attack, in the spirit of treading him down with the feet, so that he cannot rise again to the attack.

"Treading" does not simply mean treading with the feet. Tread with the body, tread with the spirit, and, of course, tread and cut with the long sword. You must achieve the spirit of not allowing the enemy to attack a second time. This is the spirit of forestalling in every sense. Once at the enemy, you should not aspire just to strike him, but to cling after the attack. You must study this deeply.

To Know Collapse

Everything can collapse. Houses, bodies, and enemies collapse when their rhythm becomes deranged.

In large-scale strategy, when the enemy starts to collapse, you must pursue him without letting the chance go. If you fail to take advantage of your enemies' collapse, they may recover.

In single combat, the enemy sometimes loses timing and collapses. If you let this opportunity pass, he may recover and not be so negligent thereafter. Fix your eye on the enemy's collapse, and chase him, attacking so that you do not let him recover. You must do this. The chasing attack is with a strong spirit. You must utterly cut the enemy down so that he does not recover his position. You must understand how to utterly cut down the enemy.

To Become the Enemy

"To become the enemy" means to think yourself in the enemy's position. In the world people tend to think of a robber trapped in a house as a fortified enemy. However, if we think of "becoming the enemy", we feel that the whole world is against us and that there is no escape. He who is shut inside is a pheasant. He who enters to arrest is a hawk. You must appreciate this.

In large-scale strategy, people are always under the impression that the enemy is strong, and so tend to become cautious. But if you have good soldiers, and if you understand the principles of strategy, and if you know how to beat the enemy, there is nothing to worry about.

In single combat also you must put yourself in the enemy's position. If you think, "Here is a master of the Way, who knows the principles of strategy", then you will surely lose. You must consider this deeply.

To Release Four Hands

"To release four hands" is used when you and the enemy are contending with the same spirit, and the issue cannot be decided. Abandon this spirit and win through an alternative resource.

In large-scale strategy, when there is a "four hands" spirit, do not give up - it is man's existence. Immediately throw away this spirit and win with a technique the enemy does not expect.

In single combat also, when we think we have fallen into the "four hands" situation, we must defeat the enemy by changing our mind and applying a suitable technique according to his condition. You must be able to judge this.

To Move the Shade

"To move the shade" is used when you cannot see the enemy's spirit.

In large-scale strategy, when you cannot see the enemy's position, indicate that you are about to attack strongly, to discover his resources. It is easy then to defeat him with a different method once you see his resources.

In single combat, if the enemy takes up a rear or side attitude of the long sword so that you cannot see his intention, make a feint attack, and the enemy will show his long sword, thinking he sees your spirit. Benefiting from what you are shown, you can win with certainty. If you are negligent you will miss the timing. Research this well.

To Hold Down a Shadow

"Holding down a shadow" is used when you can see the enemy's attacking spirit.

In large-scale strategy, when the enemy embarks on an attack, if you make a show of strongly suppressing his technique, he will change his mind. Then, altering your spirit, defeat him by forestalling him with a Void spirit.

Or, in single combat, hold down the enemy's strong intention with a suitable timing, and defeat him by forestalling him with this timing. You must study this well.

To Pass On

Many things are said to be passed on. Sleepiness can be passed on, and yawning can be passed on. Time can be passed on also.

In large-scale strategy, when the enemy is agitated and shows an inclination to rush, do not mind in the least. Make a show of complete calmness, and the enemy will be taken by this and will become relaxed. When you see that this spirit has been passed on, you can bring about the enemy's defeat by attacking strongly with a Void spirit.

In single combat, you can win by relaxing your body and spirit and then, catching on to the moment the enemy relaxes attack strongly and quickly, forestalling him.

What is known as "getting someone drunk" is similar to this. You can also infect the enemy with a bored, careless, or weak spirit. You must study this well.

To Cause Loss of Balance

Many things can cause a loss of balance. One cause is danger, another is hardship, and another is surprise. You must research this.

In large-scale strategy it is important to cause loss of balance. Attack without warning where the enemy is not expecting it, and while his spirit is undecided follow up your advantage and, having the lead, defeat him.

Or, in single combat, start by making a show of being slow, then suddenly attack strongly. Without allowing him space for breath to recover from the fluctuation of spirit, you must grasp the opportunity to win. Get the feel of this.

To Frighten

Fright often occurs, caused by the unexpected.

In large-scale strategy you can frighten the enemy not just by what you present to their eyes, but by shouting, making a small force seem large, or by threatening them from the flank without warning. These things all frighten. You can win by making best use of the enemy's frightened rhythm.

In single combat, also, you must use the advantage of taking the enemy unawares by frightening him with your body, long sword, or voice, to defeat him. You should research this well.

To Soak In

When you have come to grips and are striving together with the enemy, and you realize that you cannot advance, you "soak in" and become one with the enemy. You can win by applying a suitable technique while you are mutually entangled.

In battles involving large numbers as well as in fights with small numbers, you can often win decisively with the advantage of knowing how to "soak" into the enemy, whereas, were you to draw apart, you would lose the chance to win. Research this well.

To Injure the Corners

It is difficult to move strong things by pushing directly, so you should "injure the corners".

In large-scale strategy, it is beneficial to strike at the corners of the enemy's force. If the corners are overthrown, the spirit of the whole body will be overthrown. To defeat the enemy you must follow up the attack when the corners have fallen.

In single combat, it is easy to win once the enemy collapses. This happens when you injure the "corners" of his body, and thus weaken him. It is important to know how to do this, so you must research deeply.

To Throw into Confusion

This means making the enemy lose resolve.

In large-scale strategy we can use our troops to confuse the enemy on the field. Observing the enemy's spirit, we can make him think, "Here? There? Like that? Like this? Slow? Fast?". Victory is certain when the enemy is caught up in a rhythm which confuses his spirit.

In single combat, we can confuse the enemy by attacking with varied techniques when the chance arises. Feint a thrust or cut, or make the enemy think you are going to close with him, and when he is confused you can easily win.

This is the essence of fighting, and you must research it deeply.

The Three Shouts

The three shouts are divided thus: before, during and after. Shout according to the situation. The voice is a thing of life. We shout against fires and so on, against the wind and the waves. The voice shows energy.

In large-scale strategy, at the start of battle we shout as loudly as possible. During the fight, the voice is low-pitched, shouting out as we attack. After the contest, we shout in the wake of our victory. These are the three shouts.

In single combat, we make as if to cut and shout "Ei!" at the same time to disturb the enemy, then in the wake of our shout we cut with the long sword. We shout after we have cut down the enemy - this is to announce victory. This is called "sen go no koe" (before and after voice). We do not shout simultaneously with flourishing the long sword. We shout during the fight to get into rhythm. Research this deeply.

To Mingle

In battles, when the armies are in confrontation, attack the enemy's strong points and, when you see that they are beaten back, quickly separate and attack yet another strong point on the periphery of his force. The spirit of this is like a winding mountain path.

This is an important fighting method for one man against many. Strike down the enemies in one quarter, or drive them back, then grasp the timing and attack further strong points to right and left, as if on a winding mountain path, weighing up the enemies' disposition. When you know the enemies' level attack strongly with no trace of retreating spirit.

What is meant by "mingling" is the spirit of advancing and becoming engaged with the enemy, and not withdrawing even one step. You must understand this.

To Crush

This means to crush the enemy regarding him as being weak.

In large-scale strategy, when we see that the enemy has few men, or if he has many men but his spirit is weak and disordered, we knock the hat over his eyes, crushing him utterly. If we crush lightly, he may recover. You must learn the spirit of crushing as if with a hand-grip.

In single combat, if the enemy is less skillful than ourselves, if his rhythm is disorganized, or if he has fallen into evasive or retreating attitudes, we must crush him straight-away, with no concern for his presence and without allowing him space for breath. It is essential to crush him all at once. The primary thing is not to let him recover his position even a little. You must research this deeply.

The Mountain-Sea Change

The "mountain-sea" spirit means that it is bad to repeat the same thing several times when fighting the enemy. There may be no help but to do something twice, but do not try it a third time. If you once make an attack and fail, there is little chance of success if you use the same approach again. If you attempt a technique which you have previously tried unsuccessfully and fail yet again, then you must change your attacking method.

If the enemy thinks of the mountains, attack like the sea; and if he thinks of the sea, attack like the mountains. You must research this deeply.

To Penetrate the Depths

When we are fighting with the enemy, even when it can be seen that we can win on the surface with the benefit of the Way, if his spirit is not extinguished, he may be beaten superficially yet undefeated in spirit deep inside. With this principle of "penetrating the depths" we can destroy the enemy's spirit in its depths, demoralizing him by quickly changing our spirit. This often occurs.

Penetrating the depths means penetrating with the long sword, penetrating with the body, and penetrating with the spirit. This cannot be understood in a generalization.

Once we have crushed the enemy in the depths, there is no need to remain spirited. But otherwise we must remain spirited. If the enemy remains spirited it is difficult to crush him. You must train in penetrating the depths for large-scale strategy and also single combat.

To Renew

"To renew" applies when we are fighting with the enemy, and an entangled spirit arises where there is no possible resolution. We must abandon our efforts, think of the situation in a fresh spirit then win in the new rhythm. To renew, when we are deadlocked with the enemy, means that without changing our circumstance we change our spirit and win through a different technique.

It is necessary to consider how "to renew" also applies in large-scale strategy. Research this diligently.

Rat's Head, Ox's Neck

"Rat's head and ox's neck" means that, when we are fighting with the enemy and both he and we have become occupied with small points in an entangled spirit, we must always think of the Way of strategy as being both a rat's head and an ox's neck. Whenever we have become preoccupied with small detail, we must suddenly change into a large spirit, interchanging large with small.

This is one of the essences of strategy. It is necessary that the warrior think in this spirit in everyday life. You must not depart from this spirit in large-scale strategy nor in single combat.

The Commander Knows the Troops

"The commander knows the troops" applies everywhere in fights in my Way of strategy.

Using the wisdom of strategy, think of the enemy as your own troops. When you think in this way you can move him at will and be able to chase him around. You become the general and the enemy becomes your troops. You must master this.

To Let Go the Hilt

There are various kinds of spirit involved in letting go the hilt.

There is the spirit of winning without a sword. There is also the spirit of holding the long sword but not winning. The various methods cannot be expressed in writing. You must train well.

The Body of a Massive Rock

When you have mastered the Way of strategy, you can suddenly make your body like a rock, and ten thousand things cannot touch you. This is the body of a rock (Iwao no Mi) .

You will not be moved. This is the oral tradition. (Body like a rock means that as soon as a swordsman reaches the utmost world and is awakened spiritually; he is united with natural law. Like a rock, this law does not refer to either objects or materials. The meaning of the massive body of rock is an immovable place, an immovable mind. A mind which is free from other things, a natural, peaceful, free ranging mind).

What is recorded above is what has been constantly on my mind about Niten Ichi school sword fencing, written down as it came to me. This is the first time I have written about my technique, and the order of things is a bit confused. It is difficult to express it clearly.

This book is a spiritual guide for the man who wishes to learn the Way.

My heart has been inclined to the Way of strategy from my youth onwards. I have devoted myself to training my hand, tempering my body, and attaining the many spiritual attitudes of sword fencing. If we watch men of other schools discussing theory, and concentrating on techniques with the hands, even though they seem skillful to watch, they have not the slightest true spirit.

Of course, men who study in this way think they are training the body and spirit, but it is an obstacle to the true Way, and its bad influence remains forever. Thus the true Way of strategy is becoming decadent and dying out.

The true Way of sword fencing is the craft of defeating the enemy in a fight, and nothing other than this. If you attain and adhere to the wisdom of my strategy, you need never doubt that you will win.

The Wind Scroll

In strategy you must know the Ways of other schools, so I have written about various other traditions of strategies in this the Wind Book.

Without knowledge of the Ways of other schools, it is difficult to understand the essence of my Niten Ichi school. Looking at other schools we find some that specialize in techniques of strength using extra-long swords. Some schools study the Way of the short sword, known as kodachi. Some schools teach dexterity in large numbers of sword techniques, teaching attitudes of the sword as the "surface" and the Way as the "interior".

That none of these are the true Way I show clearly in the interior of this book - all the vices and virtues and rights and wrongs. My Niten Ichi school is different. Other schools make accomplishments their means of livelihood, growing flowers and decoratively coloring articles in order to sell them. This is definitely not the Way of strategy.

Some of the world's strategists are concerned only with sword-fencing, and limit their training to flourishing the long sword and carriage of the body. But is dexterity alone sufficient to win? This is not the essence of the Way.

I have recorded the unsatisfactory point of other schools one by one in this book. You must study these matters deeply to appreciate the benefit of my Ni To Ichi school.

Other Schools Using Extra-Long Swords

Some other schools have a liking for extra-long swords. From the point of view of my strategy these must been seen as weak schools. This is because they do not appreciate the principle of cutting the enemy by any means. Their preference is for the extra-long sword and, relying on the virtue of its length, they think to defeat the enemy from a distance.

In this world it is said, "One inch gives the hand advantage", but these are the idle words of one who does not know strategy. It shows the inferior strategy of a weak spirit that men should be dependent on the length of their sword, fighting from a distance without the benefit of strategy.

I expect there is a case for the school in question liking extra-long swords as part of its doctrine, but if we compare this to real life it is unreasonable. Surely we need not necessarily be defeated if we are using a short sword, and have no long sword?

It is difficult for these people to cut the enemy when at close quarters because of the length of the long sword. The blade path is large so the long sword is an encumbrance, and they are at a disadvantage compared to the man armed with a short companion sword.

From olden times it has been said: "Great and small go together.". So do not unconditionally dislike extra-long swords. What I dislike is the inclination towards the long sword. If we consider large-scale strategy, we can think of large forces in terms of long swords, and small forces as short swords. Cannot few men give battle against many? There are many instances of few men overcoming many.

Your strategy is of no account if when called on to fight in a confined space your heart is inclined to the long sword, or if you are in a house armed only with your companion sword. Besides, some men have not the strength of others.

In my doctrine, I dislike preconceived, narrow spirit. You must study this well.

The Strong Long Sword Spirit in Other Schools

You should not speak of strong and weak long swords. If you just wield the long sword in a strong spirit your cutting will be coarse, and if you use the sword coarsely you will have difficulty in winning.

If you are concerned with the strength of your sword, you will try to cut unreasonably strongly, and will not be able to cut at all. It is also bad to try to cut strongly when testing the sword. Whenever you cross swords with an enemy you must not think of cutting him either strongly or weakly; just think of cutting and killing him. Be intent solely upon killing the enemy. Do not try to cut strongly and, of course, do not think of cutting weakly. You should only be concerned with killing the enemy.

If you rely on strength, when you hit the enemy's sword you will inevitably hit too hard. If you do this, your own sword will be carried along as a result. Thus the saying, "The strongest hand wins", has no meaning.

In large-scale strategy, if you have a strong army and are relying on strength to win, but the enemy also has a strong army, the battle will be fierce. This is the same for both sides.

Without the correct principle the fight cannot be won.

The spirit of my school is to win through the wisdom of strategy, paying no attention to trifles. Study this well.

Use of the Shorter Long Sword in Other Schools

Using a shorter long sword is not the true Way to win.

In ancient times, tachi and katana meant long and short swords. Men of superior strength in the world can wield even a long sword lightly, so there is no case for their liking the short sword. They also make use of the length of spears and halberds. Some men use a shorter long sword with the intention of jumping in and stabbing the enemy at the unguarded moment when he flourishes his sword. This inclination is bad.

To aim for the enemy's unguarded moment is completely defensive, and undesirable at close quarters with the enemy. Furthermore, you cannot use the method of jumping inside his defense with a short sword if there are many enemies. Some men think that if they go against many enemies with a shorter long sword they can unrestrictedly frisk around cutting in sweeps, but they have to parry cuts continuously, and eventually become entangled with the enemy. This is inconsistent with the true Way of strategy.

The sure Way to win thus is to chase the enemy around in confusing manner, causing him to jump aside, with your body held strongly and straight. The same principle applies to large-scale strategy. The essence of strategy is to fall upon the enemy in large numbers and bring about his speedy downfall. By their study of strategy, people of the world get used to countering, evading and retreating as the normal thing. They become set in this habit, so can easily be paraded around by the enemy. The Way of strategy is straight and true. You must chase the enemy around and make him obey your spirit.

Other Schools with many Methods of using the Long Sword

Placing a great deal of importance on the attitudes of the long sword is a mistaken way of thinking. What is known in the world as "attitude" applies when there is no enemy. The reason is that this has been a precedent since ancient times, and there should be no such thing as "This is the modern way to do it" in dueling. You must force the enemy into inconvenient situations.

Attitudes are for situations in which you are not to be moved. That is, for garrisoning castles, battle array, and so on, showing the spirit of not being moved even by a strong assault. In the Way of dueling, however, you must always be intent upon taking the lead and attacking. Attitude is the spirit of awaiting an attack. You must appreciate this.

In duels of strategy you must move the opponent's attitude. Attack where his spirit is lax, throw him into confusion, irritate and terrify him. Take advantage of the enemy's rhythm when he is unsettled and you can win.

I dislike the defensive spirit known as "attitude". Therefore, in my Way, there is something called "Attitude-No Attitude".

In large-scale strategy we deploy our troops for battle bearing in mind our strength, observing the enemy's numbers, and noting the details of the battle field. This is at the start of the battle.

The spirit of attacking first is completely different from the spirit of being attacked. Bearing an attack well, with a strong attitude, and parrying the enemy's attack well, is like making a wall of spears and halberds. When you attack the enemy, your spirit must go to the extent of pulling the stakes out of a wall and using them as spears and halberds. You must examine this well.

Fixing the Eyes in Other Schools

Some schools maintain that the eyes should be fixed on the enemy's long sword. Some schools fix the eyes on the hands. Some fix the eyes on the face, and some fix the eyes on the feet, and so on. If you fix the eyes on these places your spirit can become confused and your strategy thwarted.

I will explain this in detail. Footballers do not fix their eyes on the ball, but by good play on the field they can perform well. When you become accustomed to something, you are not limited to the use of your eyes. People such as master musicians have the music score in front of their nose, or flourish swords in several ways when they have mastered the Way, but this does not mean that they fix their eyes on these things specifically, or that they make pointless movements of the sword. It means that they can see naturally.

In the Way of strategy, when you have fought many times you will easily be able to appraise the speed and position of the enemy's sword, and having mastery of the Way you will see the weight of his spirit. In strategy, fixing the eyes means gazing at the man's heart.

In large-scale strategy the area to watch is the enemy's strength. "Perception" and "sight" are the two methods of seeing. Perception consists of concentrating strongly on the enemy's spirit, observing the condition of the battlefield, fixing the gaze strongly, seeing the progress of the fight and the changes of advantages. This is the sure way to win.

In single combat you must not fix the eyes on the details. As I said before, if you fix your eyes on details and neglect important things, your spirit will become bewildered, and victory will escape you. Research this principle well and train diligently.

Use of the Feet in Other Schools

There are various methods of using the feet: floating foot, jumping foot, springing foot, treading foot, crow's foot, and such nimble walking methods. From the point of view of my strategy, these are all unsatisfactory.

I dislike floating foot because the feet always tend to float during the fight. The Way must be trod firmly.

Neither do I like jumping foot, because it encourages the habit of jumping, and a jumpy spirit. However much you jump, there is no real justification for it; so jumping is bad.

Springing foot causes a springing spirit which is indecisive.

Treading foot is a "waiting" method, and I especially dislike it.

Apart from these, there are various fast walking methods, such as crow's foot, and so on.

Sometimes, however, you may encounter the enemy on marshland, swampy ground, river valleys, stony ground, or narrow roads, in which situations you cannot jump or move the feet quickly.

In my strategy, the footwork does not change. I always walk as I usually do in the street. You must never lose control of your feet. According to the enemy's rhythm, move fast or slowly, adjusting you body not too much and not too little.

Carrying the feet is important also in large-scale strategy. This is because, if you attack quickly and thoughtlessly without knowing the enemy's spirit, your rhythm will become deranged and you will not be able to win. Or, if you advance too slowly, you will not be able to take advantage of the enemy's disorder, the opportunity to win will escape, and you will not be able to finish the fight quickly. You must win by seizing upon the enemy's disorder and derangement, and by not according him even a little hope of recovery. Practice this well.

Speed in Other Schools

Speed is not part of the true Way of strategy. Speed implies that things seem fast or slow, according to whether or not they are in rhythm. Whatever the Way, the master of strategy does not appear fast.

Some people can walk as fast as a hundred or a hundred and twenty miles in a day, but this does not mean that they run continuously from morning till night. Unpracticed runners may seem to have been running all day, but their performance is poor.

In the Way of dance, accomplished performers can sing while dancing, but when beginners try this they slow down and their spirit becomes busy. The "old pine tree" melody beaten on a leather drum is tranquil, but when beginners try this they slow down and their spirit becomes busy. Very skillful people can manage a fast rhythm, but it is bad to beat hurriedly. If you try to beat too quickly you will get out of time. Of course, slowness is bad. Really skillful people never get out of time, and are always deliberate, and never appear busy. From this example, the principle can be seen.

What is known as speed is especially bad in the Way of strategy. The reason for this is that depending on the place, marsh or swamp and so on, it may not be possible to move the body and legs together quickly. Still less will you be able to cut quickly if you have a long sword in this situation. If you try to cut quickly, as if using a fan or short sword, you will not actually cut even a little. You must appreciate this.

In large-scale strategy also, a fast busy spirit is undesirable. The spirit must be that of holding down a pillow, then you will not be even a little late.

When your opponent is hurrying recklessly, you must act contrarily and keep calm. You must not be influenced by the opponent. Train diligently to attain this spirit.

"Interior" and "Surface" in Other Schools

There is no "interior" nor "surface" in strategy.

The artistic accomplishments usually claim inner meaning and secret tradition, and "interior" and "gate", but in combat there is no such thing as fighting on the surface, or cutting with the interior. When I teach my

Way, I first teach by training in techniques which are easy for the pupil to understand, a doctrine which is easy to understand. I gradually endeavor to explain the deep principle, points which it is hardly possible to comprehend, according to the pupil's progress. In any event, because the way to understanding is through experience, I do not speak of "interior" and "gate".

In this world, if you go into the mountains, and decide to go deeper and yet deeper, instead you will emerge at the gate. Whatever the Way, it has an interior, and it is sometimes a good thing to point out the gate. In strategy, we cannot say what is concealed and what is revealed.

Accordingly I dislike passing on my Way through written pledges and regulations. Perceiving the ability of my pupils, I teach the direct Way, remove the bad influence of other schools, and gradually introduce them to the true Way of the warrior.

The method of teaching my strategy is with a trustworthy spirit. You must train diligently.

I have tried to record an outline of the strategy of other schools in the above nine sections. I could now continue by giving a specific account of these schools one by one, from the "gate" to the "interior", but I have intentionally not named the schools or their main points. The reason for this is that different branches of schools give different interpretations of the doctrines. In as much as men's opinions differ, so there must be differing ideas on the same matter. Thus no one man's conception is valid for any school.

I have shown the general tendencies of other schools on nine points. If we look at them from an honest viewpoint, we see that people always tend to like long swords or short swords, and become concerned with strength in both large and small matters. You can see why I do not deal with the "gates" of other schools.

In my Niten Ichi school of the long sword there is neither gate nor interior. There is no inner meaning in sword attitudes. You must simply keep your spirit true to realize the virtue of strategy.

The Scroll of Emptiness

In Buddhism, void does not imply something lacking, but rather the elimination of what is superfluous.

I will describe the essence of the Ni To Ichi Way of strategy in this book of the Void. What I call the void is where nothing exists. It is about things outside man's knowledge. Of course the void does not exist. By knowing what exist, you can know that which does not exist. That is the void.

People in this world look at things mistakenly, and think that what they do not understand must be the void. This is not the true void. It is confusion.

In the Way of strategy, also, those who study as warriors may think that whatever they cannot understand in their craft is the void. Someone like that will continue to be distracted by irrelevant things. This is not the true void.

To attain the Way of strategy as a warrior you must study fully other martial arts and not deviate even a little from the Way of the warrior. With your spirit settled on your duty, you must practice day by day, and hour by hour. Polish the twofold spirit of Shin (heart) and I (will), and sharpen the twofold gaze of ken (perception) and kan (intuition). When your spirit is not in the least confused, when the clouds of bewilderment are cleared away, there is the true void.

Until you realize the true Way, whether in Buddhism or in worldly laws, you may think that your own way is the one correct and in order. However, if we look at things objectively, in the light of the Straight Way of the Heart or in accordance with the Great Square of the World, we see various doctrines departing from the true Way. What you believe in often proves to be contrary to the true way, distorted as it is by tendencies to favor your own thoughts and views. Know this well, and try to act with forthrightness as the foundation and keep the true Heart as the Way. Enact strategy broadly, correctly and openly.

Then you will come to see things in an all-encompassing sense and, taking the void as the Way, you will see the Way as void.

In the void is virtue, and no evil. Wisdom exists, principle exists, the way exists. Spirit is Void.

(The hyoho of the body and the massive rock are both shown in Ku (emptiness). Musashi says to make your emptiness your way, and your way your emptiness. The word Ku is entirely good and contains no evil. It is a world of great wisdom which goes beyond human intellect. It benefits yourself and others, the world of Daijo (the great vehicle). It is a world in which one agrees to the true way, at any time, in any place, and agrees with many people. However, the mind presented in Gorin no Sho is the spiritual mind which frees itself from ego. Musashi called this the hyoho of Jisso-enmon (the ultimate reality and perfection), in which one receives peacefully, thankfully, everything in the world for what it is, good or bad, for the benefit of oneself and others.)

Twelfth day of the fifth month, second year of Shoho (1645)

Teruro Magonojo
SHINMEN MUSASHI

Notes

Kendo, the Way of the sword, had always been synonymous with nobility in Japan. Since the founding of the samurai class in the eighth century, the military arts had become the highest form of study, inspired by the teachings of Zen and the feeling of Shinto. Schools of Kendo born in the early Muromachi period[1] were continued through the upheavals of the formation of the peaceful Tokugawa Shogunate, and survive to this day. The education of the sons of the Tokugawa Shoguns was by means of schooling in the Chinese classics and fencing exercises. Where a Westerner might say "The pen is mightier than the sword", the Japanese would say "Bunbu Ichi", or "Pen and sword in accord". Today, prominent businessmen and political figures in Japan still practice the old traditions of the Kendo schools, preserving the forms of several hundred years ago.

---

[1] Approximately 1390 to 1600

Musashi was a ronin at the time when the samurai were formally considered to be the elite, but actually had no means of livelihood unless they owned lands and castles. Many ronin put up their swords and became artisans, but others, like Musashi, pursued the ideal of the warrior searching for enlightenment through the perilous paths of Kendo. Duels of revenge and tests of skill were commonplace, and fencing schools multiplied. Two schools especially, the Itto School and the Yagyu School, were sponsored by the Tokugawas. The Itto School provided an unbroken line of Kendo teachers, and the Yagyu school eventually became the secret police of the Tokugawa bureaucracy.

CPSIA information can be obtained
at www.ICGtesting.com
Printed in the USA
LVHW101152100522
718378LV00001B/5